$12 Million Dollar Fire at Dogwood Elementary School
Reston, Virginia

Investigated by: William A. Tobin
Hollis Stambaugh
Jennifer Roberson

This is Report 135 of the Major Fires Investigation Project conducted by Varley-Campbell and Associates, Inc./TriData Corporation under contract EMW-97-C0-0506 to the United States Fire Administration, Federal Emergency Management Agency.

Homeland Security

Department of Homeland Security
United States Fire Administration
National Fire Data Center

U.S. Fire Administration Fire Investigations Program

The U.S. Fire Administration develops reports on selected major fires throughout the country. The fires usually involve multiple deaths or a large loss of property. But the primary criterion for deciding to do a report is whether it will result in significant "lessons learned." In some cases these lessons bring to light new knowledge about fire--the effect of building construction or contents, human behavior in fire, etc. In some cases, special reports are developed to discuss events, drills, or new technologies or tactics that are of interest to the fire service.

The reports are sent to fire magazines and are distributed at National and Regional fire meetings. The reports are available on request from USFA. Announcements of their availability are published widely in fire journals and newsletters.

This body of work provides detailed information on the nature of the fire problem for policymakers who must decide on allocations of resources between fire and other pressing problems, and within the fire service to improve codes and code enforcement, training, public fire education, building technology, and other related areas.

The Fire Administration, which has no regulatory authority, sends an experienced fire investigator into a community after a major incident only after having conferred with the local fire authorities to insure that USFA's assistance and presence would be supportive and would in no way interfere with any review of the incident they are themselves conducting. The intent is not to arrive during the event or even immediately after, but rather after the dust settles, so that a complete and objective review of all the important aspects of the incident can be made. Local authorities review USFA's report while it is in draft form. The USFA investigator or team is available to local authorities should they wish to request technical assistance for their own investigation.

This report and its recommendations were developed by USFA staff and by TriData Corporation, Arlington, Virginia, its staff and consultants, who are under contract to assist the Fire Administration in carrying out the Fire Reports Program.

For additional copies of this report write to the U.S. Fire Administration, 16825 South Seton Avenue, Emmitsburg, Maryland 21727 or via USFA Web page at http://www.usfa.dhs.gov/

U.S. Fire Administration

Mission Statement

As an entity of the Department of Homeland Security, the mission of the USFA is to reduce life and economic losses due to fire and related emergencies, through leadership, advocacy, coordination, and support. We serve the Nation independently, in coordination with other Federal agencies, and in partnership with fire protection and emergency service communities. With a commitment to excellence, we provide public education, training, technology, and data initiatives.

ACKNOWLEDGMENTS

The United States Fire Administration's Major Investigation Team would like to thank the following people for their assistance with this report:

Captain David McKernan   Fairfax County Fire/Rescue Fire Investigations

Mr. Fred Ellis   Director, Office of Security and Risk Management Service
Fairfax County Public Schools

Mr. Tom Brady   Assistant Supervisor for Facility Services
Fairfax County Public Schools

Ms. Kitty Porterfield   Community Affairs
Fairfax County Public Schools

Mr. Don Greenwood   Facility Services
Fairfax County Public Schools

Captain Sam Hsu   Fire Investigator
Montgomery County Department of Fire/Rescue Services

TABLE OF CONTENTS

EXECUTIVE SUMMARY

On November 27, 2000, at 10:30 p.m., Fairfax County, Virginia, Fire and Rescue received a call from the Fairfax County Police Department reporting a fire at the Dogwood Elementary School in Reston, Virginia. The school was built in 1974 in an "open classroom" design with demountable partitions as "walls" between groups of four and six classrooms and between classrooms and halls. Measuring approximately 300' by 300', the school was a one story, noncombustible steel frame building with masonry exterior walls. Dogwood served grades kindergarten through six.

The school security monitoring station received an intrusion alarm from the north side of the school at 10:17 p.m. This was the first detected activity in a sequence of events likely related to the fire. Fire dispatch was alerted of the alarm at 10:36 p.m. and the first fire department units were dispatched at that time. Fire suppression personnel from Fairfax Fire Station 31 arrived at 10:41 p.m., within five minutes of dispatch. There were no authorized or known individuals in the school at the time of the fire; thus, no evacuation of occupants was necessary.

The fire progressed so rapidly and was burning so intensely that the first arriving fire department units quickly had to be moved and positioned further away from the building at a safe distance. During the two days of fire suppression operations, 100 firefighters and 35 units responded to the scene.

Determination of origin and cause was hampered by the extent of damage and the enormity of the dig-out, which required participation from numerous agencies and subcontractors over a two-week period of time. The school, valued at $12 million, was declared a total loss as a result of the fire. Eight months after the fire, following an intensive, thorough investigation, investigators determined the cause of the fire to be an electrical short in the building's plenum space, with the origin in Quadrant B of the structure.

INCIDENT NARRATIVE

At 10:17 p.m. on November 27, 2000, the school security central monitoring station received an annunciated intrusion alarm from Zone 7 of the Dogwood Elementary School at 12300 Glade Drive, Reston, Virginia. Zone 7 was comprised of four doors on the north (windward) corner of the building. Due to the history of false alarms at the school, the alarm was reset without investigation per normal procedure. Fairfax County School Security dispatched a security officer at 10:23 p.m., after a second intrusion alarm was received from the same zone. At the same time, Fairfax County Police were notified by school security and dispatched an officer to investigate the scene.

At 10:26 p.m., the monitoring station received a "trouble" indication from the fire panel at the elementary school. A "trouble" indication suggests an electrical fault, since there are only "pull-boxes" or "pull stations" on the school premises (no passive fire alarms or annunciated smoke detectors). At the same time (10:26 p.m.), an intrusion alarm was received at the monitoring station from an adjacent zone within the school (Zone 6). At 10:28 p.m., intrusion alarms for Zone 8, followed by Zone 5, Zone 2, and numerous other zones (9 alerts in total) were received. At 10:29 p.m., a "trouble in the phone line" alert was received.

A Fairfax County Police officer was the first to arrive at the school, at approximately 10:31 p.m., and began surveying the external premises. A second officer arrived three minutes later, and as they were walking around the back of the school (north side of the building), five or six classrooms "went off (ignited) at the same time". One of the officers immediately notified the Fairfax County Police, along with school security, of "flames at least 50 feet in the air."

Fairfax County Police and Fire Dispatch are both located in the Public Safety Communications Center, thereby minimizing any delay in sharing information. Fairfax County Fire and Rescue Department was dispatched at 10:36 p.m., after being notified by the Fairfax County Police of the report of "large flames coming from the roof" of Dogwood Elementary School. Initial dispatch included four engine companies, one rescue engine, one truck company, one tower-ladder, one ambulance, and two command units. A fifth engine company which had been dispatched for an outside fire in the area of the school shortly after the initial box assignment was added to the initial assignment once it was realized their response was related to the school fire.

Fairfax County Ambulance 431 arrived on-scene at 10:40 p.m., and immediately reported they had heavy fire showing. Engine 431 arrived at 10:41 p.m., and established a water supply by laying dual three-inch supply lines from the fire hydrant on Side One of the building to a position on Side Two. Engine 431 requested a second alarm assignment based on the heavy fire showing from the rear of the building, and assumed command of the incident per Fairfax County Standard Operating Procedure. Units dispatched on the second alarm at 10:43 p.m. included three engines, one truck, one medic unit, one ambulance, four command officers, and the canteen.

Battalion 401 arrived on scene at 10:42 p.m., assumed command, and made the decision to utilize defensive, exterior operations. All truck companies were instructed to set up for elevated stream operations, and engine companies were utilized to establish water supplies for master stream operations on all sides of the building. In addition to using master stream devices, attack hand lines and portable monitor devices were employed in exterior operations. At 10:47 p.m., ten minutes after being dispatched, fire ground units were notified that the roof of Dogwood Elementary School had collapsed on Side Three of the building.

As additional units responded, they were incorporated into exterior operations. Due to the large volume of water being used by units, a water supply sector was established. In addition to working to establish a secondary water supply, the sector officer worked with the Fairfax County Water Authority to boost water pressure in the area.

A third alarm, comprised of three engines, one truck, and an additional Battalion Chief, was requested at 11:05 p.m. Unites were directed to stage at the intersection of Laurel Glade Court and Glad Drive, approximately half a mile to the east of the incident scene. A special alarm, consisting of three engines, one tower ladder, and the command "POD" (a portable command post) was transmitted at 11:30 p.m. hours. Additional equipment was requested as necessary during the course of operations. (Appendix A, List of Responding Units).

The Rehabilitation Sector was established near the command post on Side Two of the school. Fairfax County Fire/Rescue's POD was directed to report to Side Two of the building at 12:03 a.m. (Appendix B, Fireground Map).

Building Structure and Site

The Dogwood Elementary School was located in an area of Fairfax County that is predominately residential. The school housed elementary grades kindergarten through grade six. The main entrance to the school was from the bus loop running off Glade Drive, on Side One. A secondary entrance from the parking lot was on Side Two.

Built in 1974, the school was a single story building occupying approximately 90,000 square feet. It was a noncombustible structure comprised of structural steel framing with brick and cinder block external walls. The roof had been replaced in 1994, and was constructed of corrugated steel, layers of fiberglass insulation, and five plies of roofing asphalt paper, topped with aggregate ("pea gravel").

The interior of Dogwood Elementary School was arranged in an "open classroom" design: basically a warehouse with a demountable (extruded aluminum frame) partition system for internal "walls" (Appendix C, Figures 1 and 2). Thirteen other schools in the county were constructed during the 1970's with the same design. The USFA team observed one of the schools nearby and noted that typically six to eight "classrooms" could be viewed from each hallway "wall" opening (Appendix C, Figures 3 and 4). Boundaries between classrooms consisted mostly of four-foot high bookcases, shelving, and counters and/or "soft" partitions. Boundaries between groups of classrooms were created by using demountable partitions. Additionally, the team observed that fluorescent light diffusers were stored in the plenum above the plenum access ladder at the school they visited.

The area between the building roof and the ceiling of the occupied space was approximately four feet in height, contained no significant structural impediments to air flow within the space, and was considered a "continuous, open air plenum" of approximately 360,000 feet (volume). In this configuration, the plenum was open from one end to the other (Appendix C, Figure 4) – creating an ideal environment for rapid fire extension throughout the school.

In the plenum space were the load bearing girders, bar joists, HVAC ducts (both sheet and flex), roof drainpipe (heavy gauge steel), copper plumbing feeds, and electrical wiring. No PVC tubing was observed or detected during the post-event examinations, and an interview with the school construction inspector revealed that none existed in the plenum space at the time of the fire.

Load-bearing columns were positioned 30 feet apart in the approximate north-south directions of the school and were spanned by girders. The columns were spaced 60 feet apart in the perpendicular direction and were spanned by bar joists (Appendix C, Figure 4). The columns were comprised of 8-inch square tubular (welded) wrought steel of quarter-inch wall thickness, and girders of 3/8-inch wrought steel. Seven HVAC units were positioned on the building rooftop. The HVAC unit above and closest to the probable area or origin (indicated in Appendix C, Figure 5) and similar to five of the other six HVAC units on the roof, weighted 4-1/4 tons. An aerial view of the school, both before and after the fire, can be found in Appendix C, Figures 5 and 6. Appendix C, Figure 9, shows a post-fire view of the debris under the area of the HVAC unit indicated in Figure 5.

As expected from the age range of its daytime occupants, the visit to the similar, nearby school confirmed the relatively high prevalence of plastics in the classrooms and activity rooms which, when ignited, provide high BTU-generating fuels similar to hydrocarbon accelerants. Some of the plastics in the classrooms can be seen in Appendix C, Figures 1 through 3. In terms of situational risk, the building was warehouse-like with a significant fire load of high surface area to mass ratio combustibles, many of which were high BTU-generating fuels. Compounding the fire load and contributing to substantial free burn conditions, the building constituted over a million cubic feet of available air to support combustion.

Fire Protection Systems and Equipment

Dogwood Elementary School, as well as other schools of similar construction in Fairfax County, was not subject to current code regulations since it was built in 1974. The last inspection for compliance with applicable grandfathered code was in June 2000, when a full fire alarm test was also performed.

Annunciated alarm systems, owned and maintained by the property owner, provide a local alarm system for a structure. Smoke and/or heat detectors are located throughout the structure and are wired into a central, easily accessible alarm panel that provides a visual indicator of which alarm is going off. This local alarm system is generally connected to a central monitoring station. In the event of alarm activation, the central monitoring station is alerted through a dedicated connection; they, in turn, notify local fire and police. Sprinkler systems may or may not be a part of a local alarm system. Generally speaking, automatic sprinkler systems would be installed in the ceiling of the structure, and would be activated by heat. Sprinkler heads discharge water after a glass vial or piece of metal is melted—the temperature required varies, in order that sprinklers can be installed in a wide variety of structures. Once water has begun to be discharged through a sprinkler head, a water flow alarm located on the exterior of the structure will sound, alerting passers-by. Both systems are commonly found in all types of construction. However, due to the age of Dogwood, the only fire alarm system available at the school consisted of "pull-stations" or "pull-boxes." There were no annunciated alarm systems (heat or smoke) on the premises, and no sprinkler system.

The intrusion alarms and pull-station fire alarms were in the same electrical system lines. Intrusion alarms, which were activated when a door was opened, automatically reset when the door was closed. The north side of the school is directly exposed to prevailing winds which are channeled across a playing field situated in an otherwise wooded windbreak. Wind gusts frequently triggered false intrusion alarm in the past, so a policy of automatically resetting the alarm after the first alert was established to minimize the frequency of after-hours dispatching to manually reset the alarms.

Three fire hydrants were located near the school: one 55 feet from the NW corner of the building, near the parking lot; one 120 feet from the SE corner of the building at Glade Road; and one located 200 feet from the NE corner of the school, off Laurel Glade Court.

Summary of Key Issues

Issue	Comments
Automatic fire alarms	The fire alarm system at Dogwood Elementary consisted of annunciator panels only. Therefore, when an alarm was received, it was difficult to determine what triggered it. Passive alarms may have assisted in recognizing that there was a working fire at the school, and in pinpointing the area of origin
Automatic resetting of initial intrusion alarms	The practice of resetting the intrusion alarm without investigating probably contributed to the sizes and extent of the fire. False intrusion alarms certainly are a problem in many monitored building, however, as this fire demonstrates, there can be serious consequences if fire alarms are not routinely investigated.
Lack of sprinklers	Early detection and extinguishment of a fire is critical to minimizing loss. The lack of sprinklers at Dogwood contributed to the heavy volume of fire seen by first arriving units, and to the complete devastation of the school.
Lack of firewalls and fire stops	Fire spread was unrestricted, thereby enabling the fire to rapidly and freely spread throughout the building. The presence of firewalls or fire stops could have helped to reduce fire damage and fire spread.
Large volume water supply	Due to the volume of fire and length of fire suppression operations, adequate water supply was a key factor in successful suppression. Early usage of dual water supply lines by the first arriving engine, along with good interaction between the fire and water departments, ensured that twelve master streams were supplied during fire suppression operations.
Sustained operations	Due to the magnitude of the incident, on-scene suppression operations continued unabated for two days. Personnel rotation and on-scene support, along with apparatus fueling and maintenance, were well handled during this incident.
Special Considerations	Since the incident occurred during the school year, the school district had to make immediate plans, had to relocate the students from Dogwood Elementary to other facilities for the remainder of the school year. Young students, parents, teachers, and other staff faced long travel distances to temporary quarters, which affected school, work, and family schedules. In addition, there were concerns about academic and health records that may have been lost or destroyed in the fire; roll books, along with health certificates, were originals. The immediate community and the school district collected and donated supplies and other essentials.

Automatic Fire Alarms

When Dogwood Elementary School had "pull-boxes" or "pull-stations", there were no passive fire alarms or annunciated smoke detectors. Therefore, when school security received the first trouble indication from the fire panel at the elementary school, it was assumed that there was an electrical fault. There was no way for personnel to be aware of what was occurring at the school. The lack of automatic fire alarms may have led to a delay in dispatching units to the school, as there was no clear indication of a working fire.

Automatic Resetting of Intrusion Alarms

In the minutes leading up to the fire at Dogwood Elementary School, school security had received numerous annunciated intrusion alarms. Due to the history of false intrusion alarms at the school, normal procedure dictated that the first intrusion alarm be reset without investigation. While the subsequent numerous and rapid intrusion alarms did lead to Fairfax County School Security and Fairfax County Police Officers being dispatched to the school, the procedure of automatically resetting the alarm resulted in a **nine-minute delay** that contributed to the intensity of the fire upon the arrival of School Security and Fairfax County Fire and Police Units. Not even the rapid response of these units could save the school, because the fire had gotten too great a head start.

Sprinklers

When Dogwood Elementary School was constructed, the benefit of sprinklers in schools was not fully recognized. Moreover, the cost and expense were still too high to make mandating them practical. When codes requiring sprinklers were put into place, Dogwood was not required to retrofit with sprinklers in order to be code-compliant. While sprinklers would not have prevented the fire, they certainly could have minimized the destruction caused by it. Additionally, the activation of sprinkler heads would have provided information that a fire did, in fact, exist at Dogwood.

Firewalls

Further contributing to the size of the fire at Dogwood Elementary was the lack of firewalls or fire stops in the building. As noted previously the school was essentially a large, open area with a high fire load of high surface area to mass ration combustibles, many of which were high BTU-generating fuels. There was over 1.1 million cubic feet of air available to fuel the fire—even more, if a window was open or after the roof collapsed. Firewalls could have limited the rapid growth and spread of the fire by containing it to smaller areas of the school.

Water Supply

The sheer size and intensity of the fire at Dogwood Elementary necessitated the use of ten master streams, along with several attack lines, over a prolonged period of time. Sufficient water supply and effective management were instrumental in bringing the fire under control and limiting further damage or spread to the surrounding area.

Fairfax County Fire and Rescue units are to be commended for their water supply operations on this fire. Due to the heavy volume of fire upon arrival, along with the subsequent collapse of the roof, fire suppression operations were strictly defensive and exterior. Early establishment of dual supply lines, along with placement of apparatus on all sides of the building (regardless of the presence of access roads) ensured that sustained large volume water supply operations were possible. Additionally, good coordination between the water supply sector and Fairfax County Water Authority made it possible to boost water pressure before it could become a problem and limit operations. Proactive thinking and response ensured that, although difficult logistically, water supply was ensured.

Sustained Operations

Fairfax County Fire and Rescue Units operated in a suppression capacity for two days. As a result, several logistical considerations had to be addressed to ensure that operations were sustained without interruption. A canteen was dispatched early in the incident, thereby ensuring that there was some food and beverages available on the fireground. Personnel were rotated out during the course of operations, so that no one crew was overworked. Additionally, provisions were made to refuel apparatus at the scene by utilizing Fairfax Fire and Rescue's 100-gallon fuel tanker, thereby limiting the time a unit was removed from the fireground. Per Fairfax County Fire and Rescue Standard Operating Procedure, fire units from around the county were re-assigned to cover for those on the scene.

The continued monitoring of the structure and scene safety required vigilance on the part of all personnel on-scene. High-volume water supply operations associated with ten master streams required continued attention. Late-autumn temperatures, generally in the 40's, drained the energy of personnel, both wet and dry. Personnel had to be fed, kept hydrated, and if possible, sheltered in a warm vehicle or outbuilding. Personal needs had to be addressed. All of these issues were effectively handled by Fairfax County Fire and Rescue.

Special Considerations

In interviews with school personnel, several special considerations were brought to light. The first, and most obvious, was the immediate relocation of students. The fire occurred on a Monday night in the middle of the school year, meaning that a new home had to be found for the students as quickly as possible. Children were temporarily spread among five other elementary schools—this situation lasted two weeks, or until winter break began. After the winter holiday, the Dogwood students were transferred to two smaller schools that were being closed as a result of consolidation (one was completely vacated, one was partially vacated). The students will remain at these two facilities until the new Dogwood is opened in the spring of 2002.

Another concern was the academic and health records of the students. Grades and other student records are stored both at the school and the county's mainframe computer. However, health records and emergency care cards were only kept at the school. There was initially concern that the records had been destroyed in the fire, leading parents, teachers, and administrators to attempt to gather new records for the children. Upon further salvage operations, however, it was discovered that the records had emerged mostly unscathed from the fire. The incident highlighted the need, however, for duplicate records to be kept in two separate locations.

Additionally, some teacher's grade books were lost in the fire. While many teachers took their grade books home with them at night, some others were damaged or destroyed, further emphasizing the need to maintain duplicate copies of irreplaceable records.

BACKGROUND AND INVESTIGATION

The school used no hydrocarbon fuel for eating, cooling or cooking, and was "all electric". The only hydrocarbon fuel stored on the grounds was gasoline for the lawn mowers (stored in an outbuilding), which was accounted for during the post-fire investigation. As a test for the Fairfax Fire and Rescue accelerant-detection canine, "Bert", ten dilute accelerant control samples were deposited in the fire debris and left exposed to the ambient environment (which included rain) for 10 days. Bert detected nine of the ten samples after the 10-day test, but detected no other possible trace of hydrocarbon accelerant on site other than paint thinner in the sewer system nearby. Additionally, no suspicious or questionable purchase of gasoline was detected from surveillance videos or interviews of local convenience store and gasoline dispensing station personnel.

Schools of similar construction in a neighboring county have reported serious and ongoing problems with fluorescent lighting, "flickering lights", and electrical connections. The problems, which included several electrical fires attributed to the fluorescent lighting fixtures, forced school closures until the fixtures could be replaced. This led investigators to explore the possibility that the fire could have been electrical in nature, and originated in or around one of the fluorescent lights in the school.

Based on interviews and drawings provided by teachers and staff, fire investigators could account for all major equipment (e.g., computers, printers, AV equipment, televisions) and other items in the school at the time of the fire. No controlled substances or any significant amount of money were stored on the premises.

Graffiti, consisting of the single word "ARSON", was observed spray-painted on a telephone utility box 155 feet from the driveway to the school. It was concluded by the gang unit of the Fairfax County Police that the graffiti had been present for several months at that location, and that "ARSON" was the street name of a local gang member who had spray-painted the same "name" in several other locations throughout the county. Although the graffiti was "evidence" of both state of mind and proximity, Fairfax County Fire Investigators have not established any association of the graffiti with the fire.

A number of individuals claimed credit for the fire, but all were dismissed as suspects based on subsequent interviews and follow-up by the police department and fire investigators.

Injuries and Fatalities

Upon arrival of police or fire units, there were no authorized occupants of the school, and there are no known, reported or suspected civilian injuries or fatalities. One firefighter strained his lower right back during exterior fire suppression operations.

Damage Assessment

The rapidity of propagation, due primarily to almost unlimited fuel and ventilation availability, caused such extensive damage that Dogwood Elementary was declared a total loss, valued at $12 million. The located of seven HVAC units on the roof, most weight 4 ¼ tons, combined with the overhead presence of other heavy gauge steel structure and drainpipe, significantly accelerated structural collapse.

The extent of damage, volume of debris, and enormity of the dig-out operations required thousands of extra labor hours during frigid weather and short daylight. Many leads had to be followed-up and dozens of interviews conducted. Many federal and local agencies, plus private companies, contributed to the overall investigation and clean-up; they were;

- Fairfax County Fire and Rescue Department
- Fairfax County Police Department
- Fairfax County School District—Construction Inspector and support staff
- Bureau of Alcohol, Tobacco, and Firearms
- Federal Emergency Management Agency
- Excavating and trucking operations
- Rigging equipment operations
- Insurance investigators
- Plumbing and electrical subcontractors
- Labor crews for clearing debris

- Verizon telephone crews for establishing command center lines
- Virginia Power electrical assistance

FIRE ORIGIN AND SPREAD

The school was positioned at the end of a playing field, with a natural windbreak (tree line) on the windward side of the school. Based on internal building construction and layout, HVAC induced air movement, and exposure to prevailing winds, fire propagation was fuel-controlled at all times, and probably never shifted to ventilation control.[1] Based on the fire load in the school and apparent rapidity of propagation, however, it is likely that the fire had almost unlimited access to both fuel and ventilation within a very short period of time after ignition.

Weather conditions existing at the time of the fire were: temperature 46 degrees F, dew point 35.6 degrees F, barometric pressure 29.89 in Hg, visibility 10 miles, winds 4.6 mps, with no precipitation, lightning or electric storm activity. A review of weather conditions within 24 hours before and after the fire revealed gusts of 18 to 20 mph at nearby Dulles Airport.

Post fire-interviews revealed that most of the custodial staff had departed at approximately 9:30 a.m. The last authorized individual to leave the premises was a maintenance custodian supervisor who departed at approximately 9:55 p.m. From an investigative viewpoint, it is significant that none of the custodial personnel recalled any sensory indications of characteristics suggestive of a fire already in progress prior to their departure from the school that evening. That is particularly revealing in that carbonaceous odors of combustion products would likely have quickly been dispersed or otherwise propagated throughout the "covered warehouse"-type structure through the HVAC system and open plenum.

Extremely rapid fire propagation is evident in view of the following:

- Heavy fire showing, even with the rapid response of the Fairfax County Police and Fire and Rescue units;
- "Open Warehouse" structure of the school (high ventilation access);
- Fully functional HVAC system capable of quickly drawing and dispersing large volumes of air;
- Large combustible and high BTU fire load available; and
- Complete absence of any sensory clues perceived by the custodial staff.

Determining the cause of the fire was hindered due to the enormous amount of debris that had to be sorted through. Fire origin was narrowed to a general area of the school, but a point of origin was difficult to ascertain. The first police officer on the scene observed that one of the windows on the north ("back") side of the school was broken inward, and investigation revealed the likelihood of possibly a second broken window near the first, in the north side of the school, at the time of the fire. However, no correlation could be established between the broken window(s) and subsequent

---

[1] In order to have combustion, three things must be present: heat, fuel and oxygen. These three things comprise the fire triangle. If any of these items is absent, combustion will not occur. If any one of these items is in a limited quantity, it is the controlling factor in the time and quality of the fire. Therefore, a fuel-controlled fire has unlimited quantities of oxygen and heat, and limited quantities of fuel.

events, particularly in view of the history of window replacement on the windward side (north side) of the school.

No low level fire spread indicator, V-pattern, or evidence of multiple points of origin was observed in the debris. Although numerous plant lights were typically left on 24 hours/day in classrooms in the areas of suspected origin, the working hypothesis from observations and investigation of the scene is that fire origin was high. The HVAC system, replaced in 1992 and the object of continuing maintenance and repair work since, was eliminated by fire investigators as the source or origin of the fire.

It is noted that, at a height of fourteen feet, the roofing materials in place at the time of the fire (roofing tar and asphalt paper in 5 layers) needed temperatures of only approximately 800 degrees F to become a self-sustaining "running/rolling roof deck fire."

LESSONS LEARNED

The "open classroom" style present at Dogwood Elementary School makes almost unlimited fuel and ventilation available to any fire that initiates on the premises. The absence of fire-resistant barriers or noncombustible walls between rooms containing high combustible fuel loads exacerbated extremely rapid and unimpeded lateral spread of the Dogwood Elementary School fire. Following the fire, a risk assessment was performed on other facilities of similar construction in Fairfax County. According to information released at the same time as the origin and cause findings, eleven of thirteen Fairfax County Schools built according to the same design as Dogwood will have early warning detectors installed. The remaining two schools are about to undergo renovations that include such devices.

Overhead fuels in such an open construction greatly enhance lateral extension of a fire. Although no significant high BTU plastics (other than flex ducts and possibly some stored fluorescent fixture diffusers) appear to have been present in the plenum, the roof construction (incorporating asphalt paper and tar) provided a source of overhead fuel with a relatively low endothermic to exothermic transformation temperature (800 degrees F). This feature serves to enhance overhead lateral propagation even when fire-resistant walls are in place below.

Positioning seven HVAC units weighing nearly 4-1/4 tons each on the school's roof dramatically accelerated structural collapse due to the inverse relationship of the structural steel yield stress to temperature. This phenomenon is knows as creep, where a metal can yield at stresses much below its normal yield strength when subjected to elevated temperatures (above approximately 1000 degrees F) while under load.

The presence of an annunciated passive fire alarm system incorporating smoke and heat sensors, as well as a sprinkler system, almost certainly would have allowed an earlier stage response. An earlier response may have provided an opportunity to mount an interior attack and reduce property damage, even with the "open classroom" construction design.

The policy of resetting alarms created a significant delay, allowing the fire to advance before being discovered. It is recommended that school officials find a way to reduce the number of false alarms, rather than ignoring the first one and simply resetting the alarm.

Finally, reports of electrical fixture or system difficulties in such a structure, particularly with such high occupancy, should be considered serious and resolved as immediately as is feasible.

APPENDICES

Appendix A: List of Responding Units

Appendix B: Fireground Map

Appendix C: Photographs

APPENDIX A

List of Responding Units

First Alarm:

Engines:	431,404,436,421
Trucks:	425
Tower:	436 (tower-ladder)
Rescue Engine:	425
Ambulance:	431
Command Units:	BC01, EMS 1
Additional:	E429

Second Alarm:

Engines:	415, 412, 434
Truck:	429
Medic:	425
Ambulance:	436
Command Units:	BC03, EMS 2, DFCO, SAFETY
Canteen:	2

Third Alarm:

Engines:	402, 611, 413
Truck:	403
Command Units:	BC02

Special Alarm:

Engines:	401, 423, 430
Truck:	430 (tower-ladder)
Command POD	

Command Staff:

FC	Chief Stinette
AFCO	Assistant Chief Wheatley
DCSO	Deputy Chief Martin
DFCS	Deputy Chief Rohr
DFCP	Deputy Chief Maurice
PIO	Battalion Chief Johnson

APPENDIX B

Fireground Map

(Not to Scale)

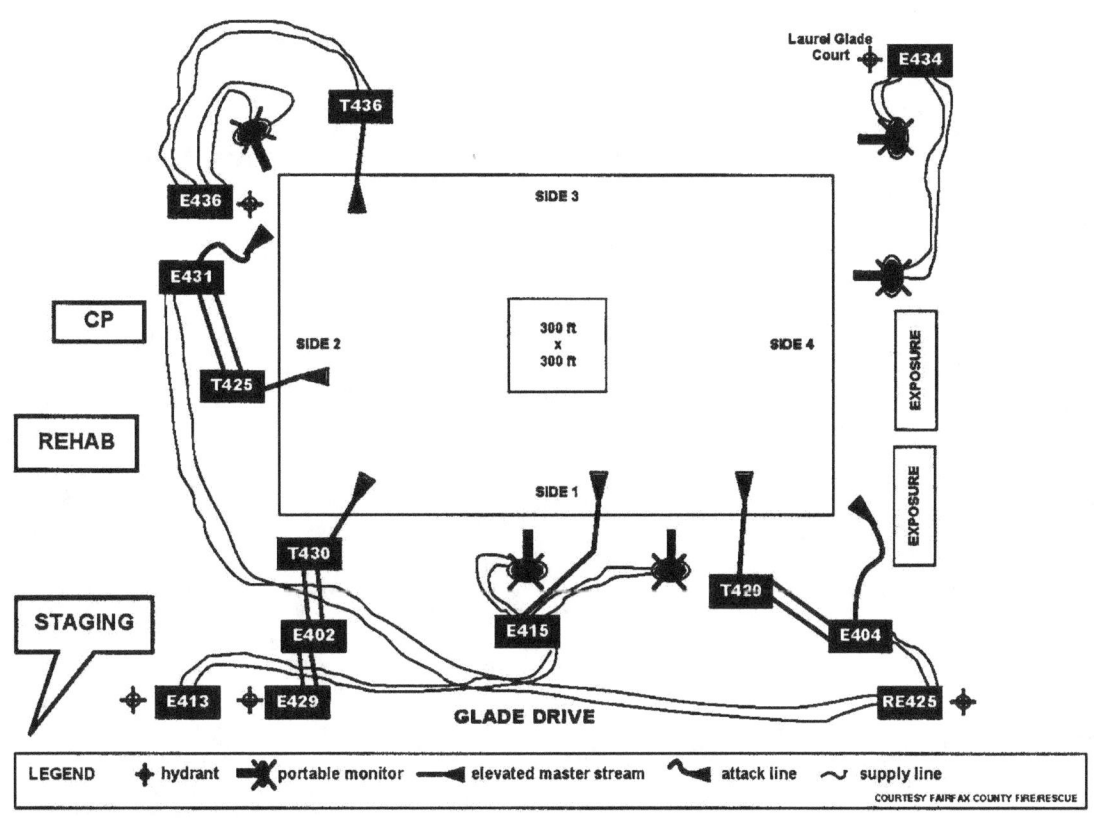

APPENDIX C

LIST OF PHOTOGRAPHS

Figures 1-2:   Views of classrooms of the "open classroom forum" style

Figure 3:   View along hallway of similarly constructed "open classroom forum" style school

Figure 4:   Similar hallway of Dogwood Elementary School after fire

Figure 5:   Aerial view of Dogwood Elementary School after fire

Figure 6:   Aerial view of Dogwood Elementary School after fire

Figure 7:   Fire suppression efforts, Sides 3 and 4

Figure 8:   Fire suppression efforts, Side 1

Figure 9:   HVAC unit after fire

Figure 10:   Collapsed roof and accordioned support column

Figure 1. View of three classrooms characteristic of the
"open classroom forum" design.

Figure 2. Library area of similar school

Figure 3. View along hallway of local school similar "open classroom forum" design. Note use of "demountable" partitions for room and hall definition.

Figure 4. Similar hallway in non-collapsed portion of Dogwood Elementary School after fire. Collapse of ceiling tiles has exposed bar joists (spanning hallway) and girders in plenum area.

Figure 5. Aerial view of Dogwood Elementary School before fire, with HVAC unit closest to fire origin indicated. HVAC unit, one of 7, weighed 4-1/4 tons.

Figure 6. Aerial view of Dogwood Elementary School after the fire.

Figure 7. View of fire suppression operations

Figure 8. View of fire suppression operations, Side 1

Figure 9. HVAC unit (arrow in Figure 5) after fire, viewed from northeast.

Figure 10. View of collapsed truss and accordioned support column
(left side of picture).